Teaching Natural Sciences, Autism and Pedagogical Practices

Daiana Tavares Chaves
Elane C. P. Padilha
Gerlany F. S. Pereira

Teaching Natural Sciences, Autism and Pedagogical Practices

Possible contributions

ScienciaScripts

Imprint

Any brand names and product names mentioned in this book are subject to trademark, brand or patent protection and are trademarks or registered trademarks of their respective holders. The use of brand names, product names, common names, trade names, product descriptions etc. even without a particular marking in this work is in no way to be construed to mean that such names may be regarded as unrestricted in respect of trademark and brand protection legislation and could thus be used by anyone.

Cover image: www.ingimage.com

This book is a translation from the original published under ISBN 978-620-2-03037-3.

Publisher:
Sciencia Scripts
is a trademark of
Dodo Books Indian Ocean Ltd. and OmniScriptum S.R.L publishing group

120 High Road, East Finchley, London, N2 9ED, United Kingdom
Str. Armeneasca 28/1, office 1, Chisinau MD-2012, Republic of Moldova, Europe

ISBN: 978-620-8-34615-7

Copyright © Daiana Tavares Chaves, Elane C. P. Padilha, Gerlany F. S. Pereira
Copyright © 2024 Dodo Books Indian Ocean Ltd. and OmniScriptum S.R.L publishing group

"Teach me in different ways, so that I can learn."
Cintia Leao Silva

"Autism can't be cured, it can be understood."
Autism Avila

"Special children, like birds, are different in their flights. All, however, are equal in their right to fly."
Jesica Del Carmen Perez

"Autism is part of this world, not a world apart."
Educating for life

"Autism: just a word. Not a sentence."
Zazzle

SUMMARY

The aim of this study was to investigate whether the pedagogical practices adopted by a natural science teacher in mainstream education contribute to the teaching and learning process in science for a student with Autism Spectrum Disorder (ASD). This is a qualitative research project, which was conducted as a case study. Data was collected through observation and questionnaires. The study was carried out in a public municipal school in the city of Macapa-Ap. The research subjects were three teachers reporting on a single student with ASD in Elementary School II: a Science teacher, a Specialized Educational Assistance teacher and an auxiliary teacher. The results were analyzed using content analysis, and showed that interaction between teachers and students with ASD is essential; that the use of differentiated and adapted methodologies in the regular classroom is necessary for the development of scientific knowledge in students with ASD; that team lesson planning is important for the scientific and civic education of these students. It follows that the results of this study confirmed the initial hypothesis about the weaknesses of science teachers who work in regular education in their pedagogical practice to work with students with ASD. In this way, they do not contribute to the process of teaching and learning science with these students. Combining this with the insufficient training of teachers, and how this negatively interferes with the learning of these students who need a truly inclusive education, we conclude that much still needs to be done to ensure that scientific literacy reaches all students, including students with ASD.

Keywords: Special education. Science Teaching. Autism.

INDICE

Chapter 1   4
Chapter 2   8
Chapter 3   16
Chapter 4   19
Chapter 5   33

1 INTRODUCTION

Teaching practices are reflected in various aspects, including the act of assessing, the act of teaching, teacher-student interaction and their relationship in the teaching and learning process. For the purposes of this text, when we talk about teaching practices, we are referring to the methodologies adopted by teachers in the classroom. These methodologies have a direct impact on teaching. Here we are particularly interested in the pedagogical practices developed in the teaching of Natural Sciences.

This teaching is considered to be of great importance to society, since science teaching based on current perspectives (such as CTS[1], for example) can train citizens who are able to think critically about everyday events and solve practical problems based on the construction of scientific knowledge. In this way, this research addresses teaching practices in science education, specifically when working with students with Autism Spectrum Disorder (ASD).

Autism is a disorder that is included in Global Developmental Disorders, according to the International Classification of Diseases - volume 10 (ICD 10), which involves a triad of symptoms characterized by extreme difficulty in communication, socialization and imagination (MINISTERIO DA SAUDE, 2013).

People with ASD generally have difficulties in developing a dialogue, some communicate through gestures and other specific forms.

---

[1] CTS - Science, Technology and Society, refers to a trend in science teaching that addresses the inseparability of science and technology from the social aspects that permeate our lives. It became known in Brazil in the 1970s, when the idea of training young researchers in science was advocated in order to keep up with the global context. Approaches involving CTS relations aim to train students to exercise citizenship and make decisions based on arguments (PEREIRA; RIBEIRO; FREITAS, 2014).

Their behavior can vary from very quiet to very agitated and, in some cases, aggressive. Some have hyperactivity associated with ASD, with the symptomatological triad mentioned above being important factors in the diagnosis of these people.

ASD was first described by Leo Kanner in 1943. Using a term from adult psychiatry, he described a group of children who had difficulty relating to people and the world around them (KLIN, 2006). As it is not considered easy to diagnose, this ends up making it difficult for people with ASD to learn, which stops inclusion in some cases, according to Praga (2011, p. 10):

> [...] it's a promising proposal, but only on paper, because in practice it proves to be very different and exclusionary. However, even in the most critical cases, it is necessary to stress the importance of social inclusion, which allows special students and other students to live together, which provides gains for everyone in the school, because they are given the opportunity to live with diversity.

Taking the above into account, it is argued that the student with ASD is a unique being, with his own characteristics, due among other things to his cognitive functions, and therefore responds to pedagogical interventions in a different, particular way and in his own time. This student needs an individualized look from the teacher and, especially, the methodologies used by the teacher. It is important to know if this student is really building scientific knowledge.

Due to numerous factors such as poor initial training and a lack of continuing training, teachers find it difficult to work more interactively with students with ASD. Caixeta et al. (2011) point out that teaching science to students with ASD has been a challenge in many ways. These include:

the lack of support and teaching materials to help the teacher and the teacher's own training (or lack of it), which ends up hindering this teaching and learning process and, in some cases, the lack of search for methods to help them.

In view of the above, the question arises: do the pedagogical practices adopted by a natural science teacher in mainstream education contribute to the process of teaching and learning science to a student with ASD?

The hypothesis raised here is that science teachers who work in regular education have weaknesses in their pedagogical practice when working with students with ASD. In this way, they do not contribute to the process of teaching and learning science with these students. This teacher should train scientifically literate citizens .[2]

The aim of this research was to investigate whether the pedagogical practices adopted by a natural science teacher in mainstream education contribute to the process of teaching and learning science to a student with ASD.

The literature shows that people with ASD have qualitative alterations in social interactions, communication, and a restricted, stereotyped and repetitive repertoire of interests and activities. Consequently, they also have learning difficulties, which may or may not be associated with limitations in the process of biopsychosocial development. This is why many consider it a real "revolution" for people with ASD to have the right to education and to be educated in mainstream schools.

---

[2] Scientific literacy is one of the main objectives of science teaching, according to Chassot (2014). The author states that scientific literacy is the body of knowledge that would make it easier for men and womento read the world in which they live [...] it would be desirable for those who are scientifically literate not only to be able to read the world in which they live, but also to understand the need to transform it, and to transform it for the better (CHASSOT, 2001, p. 38).

But despite all the progress made in terms of legislation with the creation of Law No. 12.764, of December 27, 2012, which guarantees the National Policy for the Protection of the Rights of People with ASD and establishes guidelines for its implementation, as a result of various international meetings in favor of the rights of these people, we still see an educational framework that urgently needs improvement. Despite the existence of a legal framework in our country that supports and sustains education, there is still a great need to improve the quality of education.

From this perspective, we believe that the research proposed here is important because of the need to understand the pedagogical practices adopted by science teachers in regular education and, consequently, how they interfere in the teaching and learning process of these students.

2 LITERATURE REVIEW

2.1 Autism Spectrum Disorder (ASD): Brief Considerations

Autism is a disorder that belongs to Global Developmental Disorders (GDD) (PASSERINO, 2012). The literature on autism highlights as a relevant aspect in its identification a triad (WING, 1998) based on the elements of social interaction, communication and language and behavior (RIVIERE, 2001). These elements have qualitative characteristics that make them peculiar or *deficient* in the case of autism (PASSERINO, 2012).

Smith (2008) put together a diagnostic framework with some behavioral characteristics, including that people with ASD don't flinch, don't ask for a lap or protection when they get hurt, avoid maintaining physical and visual contact and remain indifferent to the people around them. The causes of the disorder are inconclusive. As Mello (2004, p. 20) points out:

> It is believed that its origin lies in abnormalities in some part of the brain that has not yet been conclusively defined and is probably of genetic origin. It is also thought that it could be caused by problems related to events that occurred during pregnancy or at the time of delivery. The hypothesis of an origin related to maternal coldness or rejection has been ruled out and relegated to the category of myth for decades.

According to the *Diagnostic and Statistical Manual of Mental Disorders* fourth edition (DSM-IV), published by the American Psychiatric Association (2000) for the classification of mental disorders, and used to diagnose autism spectrum disorder, autism can be subdivided into five groups: I - classic autism, II - Asperger's syndrome, III - disintegrative disorder, IV - GDD - global developmental disorder - unspecified. The following is a brief understanding by authors of the different types of autism.

According to Pessoa (2013) in classic autism, sufferers can range from people with above-average intelligence and verbal skills to those with cognitive alterations and a complete lack of spoken language.

Asperger's Syndrome is a type of high-functioning autism. In it, people have some problems in the social area, but do not have any global delay or retardation in the cognitive development of language (PERORAZIO, 2009).

With regard to Childhood Disintegrative Disorder, Gillberg (2005) states that

> This extremely rare disorder occurs when there is normal development until the age of 3 or 4 and then regression occurs, which is sometimes due to underlying neurometabolic disorders. But at our current level of understanding of neurometabolic disorders, there are no good clues as to what mechanisms are involved in their pathogenesis.

As ASD is not a disorder of known origin, this makes diagnosis difficult. Mota and Sena (2013, p. 3) point out that "[...] its diagnosis is made by the specialist Neurologist, together with a multidisciplinary team, with Psychologist, specialized Psychiatrist and, with the use and functionality of specific criteria". In this case, the clinical picture is a decisive factor in determining the diagnosis.

For Assumpgao Junior and Pimentel (2000), due to the characteristics that some people with ASD have, it is necessary in most cases to use medication. For the author, this is complex, focusing on the reduction of target symptoms, represented mainly by agitation, aggression and irritability, which prevent these people from being referred to stimulation and educational programs.

These students need pedagogical and institutional support in general to make their development possible. Thus, their attention, coordination, memorization and other aspects must be stimulated.

In these terms, Milagre and Souza (2011) point out that children with ASD are no longer seen as "sick", but as subjects with potential to be developed in the pedagogical field and in other areas of their lives.

It is therefore important that the pedagogical resources and practices applied to these people are differentiated and enable their full development.

2.2 Pedagogical practices and science teaching

For Sacristan (1999), pedagogical practice is understood as the action of the teacher in the classroom. In fact, this text shares this thought, adding that these practices reflect multiple dimensions, including assessment, student-teacher interaction and the teaching and learning process itself. It also refers to methodological issues, which are the focus of the analysis of teaching practice in science education, for the purposes of this research.

Other perspectives on teaching practice are presented in the literature, such as the one defended by Tozetto and Gomes (2009, p. 9) "The practice of the teacher as a social and cultural agent should help the student to overcome obstacles in the construction of their knowledge". For Pimenta (2002), the Marxist perspective, for example, involves knowledge of the object, establishing goals and intervening in the object so that reality can be transformed as a social reality.

The author thus points out that the intentionality that Marx proposes will give practice a different character from the practice present in the concept of technical rationality. The intention and action to transform reality, present in practice, give this human activity the relationship between theory and practice for the transformation of nature and society, in other words, praxis (PIMENTA, 2002).

What is needed here is a teacher with the knowledge to deepen the student's understanding. In this sense, their work cannot be mechanical; on the contrary, it must be interactive and take place in a dialectical movement, based on the student's knowledge.

Based on this thinking, Pereira and Padilha (2014a; 2014b) defend the need for science teaching to build scientific knowledge along the lines of Ausubel's and allude to the importance of students' prior knowledge, called subsungores, for building scientific knowledge.

Sacristan (1999) mentions that the teacher takes on the role of reflective guide, i.e. the one who sheds light on classroom actions and significantly interferes in the construction of the student's knowledge. Thus, in carrying out this task, he or she provides reflections on pedagogical practice, because it is based on the assumption that, in assuming the problematizing attitude of practice, one modifies and is modified, generating an objective culture of educational practice. He also argues that "[...] educational practice is the end product from which professionals acquire practical knowledge that they can improve" (SACRISTAN, 1999, p. 73).

In these terms, it is argued here that the practices to be developed by teachers will guide them in the sense of which teaching methods should be most satisfactory for the development of content to be taught in the classroom. In particular, science content stands out here.

What can be observed is that there is a great lack in the activities carried out in everyday school life with regard to mastery of the knowledge to be taught (especially specific science content), because teachers have been shown to be unprepared to carry out the task of teaching, either because of a gap in their personal knowledge or because of a hybrid education.

In the case of teachers who are going to work with students with ASD, this hybrid training is even more necessary, because if it is already challenging to teach science to so-called "normal" students, imagine how much more complex it is to develop scientifically literate students with ASD.

> With inclusive practice, teachers are not held hostage by any kind of instrument that labels and marginalizes students. They are free to discover their students in everyday school life. Teachers know that, in order to teach, it is necessary to get to know each student, to get close to them, to discover with them the best paths, their learning style, their pace, their needs, their possibilities. They also know that no clinical care, diagnosis or examination carried out in a strange and impersonal environment will be able to provide real, meaningful and permanent information about a child or young person (SANTIAGO, 2004, p. 24).

Inclusive practice takes place through interaction with students. In this way, the teacher can implement methodologies for their development. Education is a key factor in the educational process of students with ASD.

It is worth pointing out that, in order to educate a person with this disorder, it is also necessary to promote their social integration and, at this point, the school is a great mediator for this integration to take place, making it possible for them to acquire important concepts for the rest of their lives.

> If teaching methods are used correctly, they are an excellent tool to benefit learning. Teaching methods are, therefore, ways of presenting a particular topic or subject in such a way as to make learning both efficient and enjoyable (AYRES, 2011, p. 95).

Educators constantly need to be reviewing their practice in a close interlocution with the knowledge that emerges with a view to their personal and professional development through the quality of education they intend to offer their students (GUEDES; SANTANA, 2011, p. 31).

According to Gomes (2009), teachers are anxious or reluctant about their educational inclusion in the face of the difficulties that people with ASD have with communication and behavior resulting from neurological or psychological conditions.

In these cases, the school has a fundamental role, not only to include these students in the school environment, but also to monitor their performance, since all students with the disorder have the right guaranteed by Law No. 12,764 of December 27, 2012 (BRASIL, 2012) to attend regular school classes.

In addition, according to Paragraph 1, "In cases of proven need, the person with autism spectrum disorder included in ordinary regular education classes, under the terms of item IV of art. 2º , will have the right to a specialized companion" (BRASIL, 2012).

People with ASD should not be seen as "incapable of learning", but rather as people who have different ways of achieving this learning. According to Law No. 12.764, which guarantees them the right to be treated as disabled, according to paragraph 2, "A person with autism spectrum disorder is considered a person with a disability for all legal purposes" (BRASIL, 2012). The right to be treated as disabled guarantees the right to the country's inclusion law.

Inclusive Education, on the other hand, implies that the person with ASD is no longer expected to fit in with the so-called normal students. The aim is for them to reach their full potential alongside their "normal" peers. The school environment will play a role in the development of these children.

According to Caixeta et al. (2011), inclusive schooling requires schools to review their teaching practices: from the content taught in the classroom to the methodology employed by teachers, since it is based on the ideal that different students require different resources and mediation strategies.

For the teacher, it's a little more complicated, as there is tension and expectation regarding the student's performance in the classroom, not only on the part of the family, but also on his own part, as he wants to make that child surpass himself.

According to Passos (2008), the idea of working with the 'different' leads education professionals to look for a large number of activities and strategies to keep them busy, which ends up compromising the natural development of these children's abilities.

In this way, if the teacher works dynamically with the students, science teaching has a lot to contribute not only to their knowledge of the subjects taught, but also to their understanding of everyday things, in the terms of Cachapuz et al. (2011).

The importance of studying science is mainly due to the fact that it enables people to develop a critical view of the reality that surrounds them, so that they can use the knowledge they acquire in everyday life, analyze different situations and be able to evaluate issues of importance in determining their quality of life.

In the specific case of science teaching, it is necessary to adapt pedagogical strategies, teaching materials, games and games so that the mediation of science content is improved in view of the characteristics of these students, thus enabling scientific literacy.

3 METHODOLOGY

3.1 Characterization of the study area

This research was carried out at the Hildemar Maia Municipal Elementary School in the municipality of Macapa, located at Avenida Conego Domingos Maltes n° 52, Bairro do Trem. The school offers regular primary and secondary education from an inclusive perspective, and is considered to be the first inclusive school in the municipality, catering for several students with some kind of Specific Educational Need (SEN). The school has 10 students with ASD in Primary I and one student in Primary II.

In terms of infrastructure, the building is divided into two blocks, containing 8 classrooms, a boardroom, a teachers' room, a computer lab, a multifunctional resource room for Specialized Educational Assistance (AEE), an indoor sports court, a kitchen, a library, a bathroom inside the building, adequate toilets and facilities for students with disabilities or reduced mobility, a secretarial room, a cafeteria, a pantry, a multimedia room, a front garden, a vegetable garden, a sensory garden and filtered water, mains water, artesian well water, mains electricity, a cesspit, garbage for periodic collection and Internet access.

3.2 Data collection and analysis

This is a study with a qualitative approach, using a case study to contribute to a better understanding of individual phenomena. According to Yin (2001, p. 27). "The case study relies on many of the techniques used in historical research, but adds two sources of evidence that are not usually included in a historian's repertoire: direct observation and a systematic series of interviews."

To collect the data, face-to-face observations were carried out during Natural Science lessons in the elementary school class in which the student with ASD was enrolled. To do this, we followed an Observation Guide (Appendix A) about the methodology used by the teacher to work on science content with students with ASD.

According to Yin (2001, p. 115) "Observational evidence is generally useful for providing additional information on the topic being studied". There was a two-month period of observation, which took place three days a week, on Mondays, Wednesdays and Fridays. The observations took place in May and June 2015, due to the strike in April 2015. Interviews were also carried out with the teachers (Appendix B), in order to assess the reality found at the school.

The data was analyzed using content analysis, according to Bardin (2011). The "Analysis of answers and open questions" proposed by the author in question was used as a reference. This type of analysis is more abundant with information, as it takes into account the interviewee's view, with the aim of valuing and treating the original textual information as automatically as possible, without transformation or "coding", or "*a priori* reworking of the basic information" (BARDIN, 2011).

3.3 Research subjects

The subjects of the study were the school's Science teacher, here called Science Teacher (PC), as well as the AEE teacher, here alluded to as (PAEE) and the Discipline Assistant Teacher (PAD).

The student with ASD, referred to here as ATEA, was the object of indirect observation, but this was crucial to the development of this study.

We only observed the science classes that took place in class 622 6th grade, elementary II, where there was a student with ASD. These observations took place over two months on Mondays, Wednesdays and Fridays.

3.4 Ethical aspects of research

With regard to ethical aspects, as this study is qualitative in nature, it will not be subject to evaluation by the Research Ethics Committee in accordance with Resolution No. 466/2012, which deals with research and tests on human beings.

However, all the established ethical precepts will be respected with regard to ensuring the legitimacy of the information, privacy and confidentiality of the information, when necessary, making the results of this research public.

The people who took part in this research have a degree in Biological Sciences, the other two in Pedagogy, one with a Specialization in Special Education. They allowed us to take part in the research and collaborated whenever necessary, passing on any information that was pertinent to the work being carried out and relating their experiences of everyday school life.

The Free and Informed Consent form (attached) was signed.
A) and the Consent of Participation of the Person as Subject (annex B)

4 RESULTS AND DISCUSSIONS

Before the actual observations, there were preliminary informal conversations with the teachers taking part in the study. In this conversation, the CP reported that her interaction with the ATEA had a 'deficit' due to a lack of in-depth knowledge about ASD, and that she had already done some superficial research into it and some methodologies for working with it in the classroom. But, according to her, without success. She reported that she interacted little with the ATEA.

Farias et al. (2008, p. 24) point out that:

> [...] teacher-student interaction is understood as an interactive two-way process, in which the mediator must involve the mediated in problem-solving activities in order to help them face the tasks/problems effectively and enrich their behavior with a view to the student's autonomy in future activities.

The PAEE was optimistic about the development of the ATEA, but had difficulties with the other teachers who couldn't understand the change of routine for the ATEA, *"who until then had only one teacher in the class and now have to adapt to a change of timetable with several teachers coming in and out of the classroom".*

The PAEE also reported that teachers have a hard time understanding the function of the AEE, since from their point of view it is a reinforcement room. According to Fortes and Bridis (2006), what prevents mainstream teachers from working effectively with students with ASD is the fact that they..:

> They experience a certain "fear" of working with people with autism, either because of their lack of knowledge about the autistic condition itself or because they are

faced daily with the possibility of not being able to find answers when they intervene pedagogically with such a student (FORTES; BRIDIS, 2006, p. 28).

In this perspective, they rely on AEE teachers, and also due to a lack of knowledge about the role of these professionals, which according to the Secretariat of Special Education (BRASIL, 2008) is to identify, develop and organize pedagogical and accessibility resources that eliminate barriers to the full participation of students, taking into account their specific needs.

The PAD previously reported that the teachers 'think' she's there to teach the student, not adapting the material given in class.

Peres de Paula (2014, p. 13) alludes that

> The adaptation of didactic-pedagogical resources and materials is essential in this regard, not only enriching classroom work, but also enabling teachers to work individually with each of their students.

In this way, it can be seen that the teachers do not understand the role of the assistant teacher, which according to SE Resolution No. 02, of 12-12012 is:

> Article 4 - The Auxiliary Teacher, referred to in item I of the previous article, will have the primary function of supporting the teacher responsible for the class or subject in the development of teaching and learning activities, especially those of continuous recovery, offered to students in primary and secondary education, with a view to overcoming difficulties and needs identified in their school career.

The following are some considerations about the observations made during the collection of data for this study.

4.1 The Observances

During the observation period, in the first week, the CP was not very perceptive to the presence of the ATEA in class, only mentioning him during roll call and not interacting with the student at any time. After the second week, PC and PAEE met to plan and it was agreed that PC would use new methodologies in class such as videos, posters and films with the help of PAEE in order to facilitate the development of the student with ASD as well as their learning.

Pinheiro and Costa (2015, p. 15) allude that,

> There needs to be a professional pedagogical relationship between the regular classroom teacher and the ESA teacher, where the regular classroom teacher's role in the face of inclusion is to activate the ESA teacher and maintain a close relationship. Their role is to accompany students with SEN, work in partnership with the ESA teacher, organize activities for the students, talk about the performance of students with special educational needs.

One of the PAEE's suggestions was to use videos about the content studied in class. This was well accepted by the PC. Three videos were shown explaining the formation of the Earth. In the first one, which was a little longer at around 8 minutes, the ATEA paid close attention and, as the video was subtitled in English and Portuguese, he kept looking at the screen the whole time.

In the second video, which was narrated entirely in Portuguese, the ATEA remained attentive for a few minutes and then lay down on the floor until the end of the third video, after which he left the room and refused to return. The other videos were 3 and 5 minutes long. The teacher reinforced the subject of the videos by asking questions orally so that the class could interact and clear up any doubts, but ATEA didn't participate.

The literature shows that working with cinema or other forms of media can make science teaching more interesting (PEREIRA; PADILHA; COSTA, 2015). What's more, if it's done properly, always linking the movie or video in question to the subject being studied in class.

It's necessary to look for videos covering content that catches the student's attention and the teacher needs to mediate the visual content of the video with the written content of the textbook.

In one of the classes that was held outside the classroom (because it had no environmental conditions due to refrigeration problems), the PC taught content from the textbook and only halfway through the lesson did she realize that the ATEA was in the school, because he had left the computer room where he was with the PAD doing science activities.

It can be seen from this that the CP's degree of inattention is so great that she didn't notice the student's presence at school, and at no point did she ask the students in the class if the ATEA had been absent. The literature shows that the effective participation of teachers in the teaching and learning process is essential for the development of students with ASD,

> [...] to better understand the importance of the teacher in the development and inclusion of children with disabilities, especially those with autism, we can adopt the concept of ZDP (Zone of Proximal Development), understood as a domain of constant transformation based on the maturation of cognitive functions that will be consolidated in the future (OLIVEIRA, 1997, p. 23).

However, what we saw was that the PC did not seek to get closer to the student in an effective way, leaving only the PAD with responsibility for the student with ASD in class, even during his lessons.

Bearing in mind that ATEA didn't go to school every day, whenever he did he interacted a lot with a particular classmate and more with the girls in the class than with the boys. We noticed that the CP hardly mentioned her name.

On a day when she was taking notes, she was surprised to see for the first time ATEA's notebook, which was full of activities she had written down and adapted by PAD; she saw his handwriting and was surprised to learn that he was doing the activities. This showed that there was a total lack of concern on the part of the CP about the development of ATEA.

According to the Guidance for Teachers (2014, p. 30).

> Appropriate behavior should be positively reinforced and inappropriate behavior should NOT be reinforced. When inappropriate behavior occurs, the appropriate behavior that is expected should be taught. Repeat the teaching as many times as necessary [...].

When the PAD entered the class, the student was agitated and crying because he had been refused pens. As he was nervous, she sat him down at his desk, gave him several pieces of paper and pens and encouraged him to draw. Within a few minutes he had already drawn three pictures and, after that, he seemed much calmer and concentrated on drawing. But he stopped and looked at what he had already drawn, and in each of the drawings he put a word in English.

The PAD's attitude of calming down the ATEA, encouraging them to do the right thing.

that he likes is not only effective, but also develops the student's other skills, because by drawing he can train his motor coordination and, because he knows English and likes English, he can practice writing in the language in question.

PC brought the whole class an adapted piece of work in which there were various pictures and the students had to identify the names of the images in the appropriate spaces. The main figure was about the Earth and its components. ATEA, with the help of PAD, did half of the activity, refusing to finish and leaving the room. PAD made a note in his notebook for his mother to help him at home.

ATEA has difficulty staying in class for long, so much so that in his classroom in particular, there is a latch at the top to prevent him from leaving during lessons, as he loves to "wander around" the school, especially in the sensory garden.

Students in the class are divided into groups, but everyone tries to interact with the ATEA and the PC is more open to the ATEA's mood swings, socializing with his PAD on activities for him to carry out. In this harmonious atmosphere, the bimester ended and everyone went on holiday.

4.2 The Questionnaire

The first question was as follows: "During your undergraduate studies, were you prepared to know how to teach classes and how to deal with students with autism spectrum disorder?".

PC and PAD answered *"No"*. The PAEE mentioned that *"Yes and no. We only had the theory"*. This shows that undergraduate courses, both in the past and today, are not very concerned with Inclusive Education, and in the case of PC and PAD, not even just the theoretical part of the subject. Inclusive education occurs very frequently in speeches, but little reverberates in practice. According to Alves (2002, p. 7)

Inclusive education presupposes education for all, not just in terms of quantity, but also quality. This means that students must be able to appropriate both the knowledge available in the world and the forms and possibilities of new productions for creative insertion into the world.

The second question in the questionnaire was more specific, namely: "What practices have you adopted for the development and integration of students in the classroom, especially those with autism spectrum disorder?".

The answers were very varied. The CP mentioned that *"In the case of the autistic student, the teacher who accompanied him adapted the activities"*.

The PAD replied that she did this through *"Conversations, dialog, movies and various games, etc. [...]'*.

The PAEE mentioned that it was done through *"Planning work in partnership with the class teacher. Planning to achieve the whole"*.

This shows that the PAD really does play a crucial role in this student's development, since the CP left all responsibility for adapted material to this teacher and to the PAEE, when in her speech she gives the impression that she is doing everything correctly, when in fact she is not.

The third question was "What is your greatest difficulty with students with autism spectrum disorder included in mainstream classes?".

The PC reported that there was *"a lack of dialogue, as the student didn't express himself in any attempt to communicate"*.

The PAD replied *"when teachers don't plan"*, and the PAEE said it was *"silence and interaction"*.

The answers pointed to different difficulties regarding the presence of a student with ASD in class. The CP who reported the student's lack of communication, making clear a misunderstanding about ASD, which is a characteristic difficulty for those with the disorder. However, it can be overcome as long as there is encouragement to find an appropriate way of interacting with the ASD. Despite having a 'limited' verbal vocabulary in the classroom, he was able to interact with the rest of the class through gestures and attitudes.

According to Bereohff (1991, p. 7) "[...] in order to educate an autistic child, it is necessary to take into account the lack of interaction with the group, poor communication, speech difficulties and the changes in behavior that these children display".

The PAD says that the fact that teachers don't plan ahead makes it difficult for the student to take part in some classes properly, because she is the one who adapts the activities. While, for the PAEE, it's the silence of the other students that bothers a person with ASD because their ears are sensitive to noise.

As well as the ATEA, which often makes stereotyped sounds and speaks loudly or even shouts in class at times, and interaction, because for the ATEA to interact during the lesson in relation to the content, even to do the activities, only if it was linked to the image of the Mickey Mouse® cartoon characters, which was done through stickers.

The fourth question was "Do you use any teaching resources to help you understand the content better?".

PC mentioned *"Datashow (slide show), poster making and presentations"*.

The PAD replied, *"Yes, junk, games, magazines, cutting and pasting"*.

And the PAEE reported that she uses *"educational software, folders, and adapted material".*

PC reported using *slides,* but during the observations, it was noted that she needed to adapt her methodology with more explanatory images and make them more appealing to the ATEA. Meanwhile, PAD uses collages, which have proved to be very conducive to the understanding of the student in question, since he likes to draw. Abujadi (2014) describes that:

> [...] patients with autism have preference networks, which make them organize themselves. I have to link the new information of the day to these neural networks of interest. So if a patient is very interested in animals, in drawing pictures, and I want them to be interested in other things, I can use the animals to give them knowledge about other areas. (video)

In this way, it is clear that each being is unique, and although some behavioral characteristics are similar, each one has a particularity that must be respected.

The material that the PAEE uses in the AEE helps the student with ASD to develop their difficulties by exercising reading, writing and their level of concentration to complete tasks, among other things.

The fifth question was: "Is this student managing to develop didactically in the school environment? And how does this happen?".

The CP mentioned *"I don't see much change during the (student) school term".*

The PAD replied *"I was succeeding and suddenly there was no teamwork".*

The PAEE said*:* "Yes, because the school has an inclusive vision, and they work on the inclusive theme from day one. The student feels included, making the process easier".

In this question, the teachers answered differently, seeming to be talking about different students, when in reality they were answering about ATEA.

The PAD, having already been ATEA's head teacher during the 4th and 5th grades, mentioned that he was able to develop, but that the lack of teamwork, where the adaptation of one teacher in the classroom to several and the change of timetable hindered this development. However, In the PAEE's opinion, he was managing to develop because he felt welcome in the school and without distinction from the other pupils.

The sixth question was "In relation to science teaching, how is it suitable/adapted/developed for this student?"

PC mentioned that it was *"adapted by me and the teacher (caregiver)"*.

The PAD replied that it was *"through cutting and pasting classes"*.

The PAEE reported that through *"adapted handouts, videos and folders"*.

Science teaching should be more dynamic and interactive, because many of the activities worked on in the theories set out in textbooks can be done in practice, through experiments, practical lessons, and countless other ways of making learning more interesting for the student. Souza (2013) points out that in the context of science teaching, this is exactly the challenge that the vast majority of educators are experiencing at the moment: including those excluded from the global educational process, having to deal with differences and their uncertainties.

The seventh question was: "During science lessons, do students show any interest in the content?"

PC answered *"None"*.

PAEE simply mentioned *"Yes"*.

While the PAD said: *"Yes, I like it very much, but the mediator needs to take care to develop activities that the student likes"*.

In this question, the PC's answer showed that in her view the student was not interested at all in the classes, while the PAEE and PAD were more positive than yes, and the PAD also informed of the student's interest, identifying a lack of greater interest on the part of the teacher to develop activities for the ATEA.

Science teaching has required teachers and students to understand a new world context in order to meet the demands of today's society. "In these terms, this teaching must be consolidated in order to help prepare students so that they are able to intervene effectively and actively in the society in which they are inserted" (PEREIRA; RIBEIRO; FREITAS, 2016).

To this end, Leon (2003) mentions that it is important for contemporary individuals to be equipped with scientific training, skills and attitudes that can equip them, throughout their lives, to take a critical, autonomous stance and make decisions, in line with current social aspirations.

"In this way, we highlight the importance of bringing alternative perspectives to science teaching. These can be consolidated in such a way as to instigate in students the autonomous search for knowledge" (PEREIRA, RIBEIRO; FREITAS, 2016).

According to Sadler (2004), effective science and technology education requires the active involvement of students in research of this nature, and the development of discursive practices that enable them to apply their scientific knowledge to personal and public decision-making.

But for this to be possible, it is necessary to include the social aspects of science and technology in science teaching. This will certainly help prepare students for future scientific and technological challenges (PEREIRA, 2012; PEREIRA; RIBEIRO; FREITAS, 2014).

In the same sense, in order for there to be concrete social participation in these aspects, it is essential to have an adequate scientific education that makes it plausible to understand the existing problems, as well as the options for solving these problems (GIL-PEREZ; VILCHES, 2006).

Thus, we agree that "Increasing the level of public understanding of science is now a necessity, not only as an intellectual pleasure, but also as a need for human survival" (LORENZETTI; DELIZOICOV, 2001, p. 5).

In these terms, it has become "[...] a cultural necessity to broaden the universe of scientific knowledge, given that today we live more intensely with Science, Technology and their artifacts" (LORENZETTI; DELIZOICOV, 2001, p. 5).

Today, there are countless methodologies that natural science teachers can use in the classroom to achieve these aspects of teaching. It is understood here that promoting scientific education is the teacher's responsibility, so that critical people who understand the current transformations taking place in the world emerge from the classroom.

We advocate playfulness, the use of the most varied artistic expressions (dance/body expression, photography/imaging[3], theater, cinema[4], music[5], poetry, sculptures - these are already present in recent

---

[3] Santos et al. (2015) carried out a study aimed at understanding students' impressions of the experience of photographing socio-environmental issues and the possibilities of using them in science teaching.

[4] This study aimed to discuss the relationship between the movie The **Croods®** and science teaching. In addition to presenting other conceptual possibilities to be addressed in science classes (PEREIRA; COSTA; PADILHA, 2015).

[5] The aim of this study was to enable students to relate their spontaneous knowledge of waves and acoustics to the teaching of natural sciences in the classroom, covering subjects such as music and musical instruments, waves and acoustics in a dynamic,

publications in science teaching), and especially experimental practices. These should not be dissociated from science teaching in elementary school, as they enable learning that is meaningful to the student, in the terms of Pereira and Padilha (2014a, 2014b) .[6]

The eighth question was "How is your relationship with the ESA teachers, and do they give you tips on suitable methodologies?"

Asking the same question, however, was changed to the ESA teachers as follows: "How is your relationship with the student's class teachers, and do they seek advice on appropriate methodologies?"

The PC answered *"Yes, sometimes"*.

PAD simply mentioned "*Sometimes"*.

And the PAEE reported that *"Yes, this interaction and socialization is fundamental for new practices"*.

The CP's response shows that the communication between her and the professionals in the ESA room is not satisfactory, because sometimes it doesn't represent that their relationship is one of partnership as it should be.

There is a great lack of interest on the part of CP in the teaching and learning process of ATEA Science. This is very common in schools in general. This interaction would be essential, since this process is essential to the development of both "special" and "normal" students.

The ninth question was: "What content did you teach in class that you felt was easier for the student to assimilate?"

For PC: *"None"*.

The PAD replied: *"Portuguese, Science and Geography"*.

---

constructive and innovative way with the help and encouragement of the viola for the study of physics in elementary school (FURTADO; KAWAKAMI JUNIOR; PEREIRA, 2015).

[6] In their studies, the authors emphasize that David Ausubel's Meaningful Learning Theory (SLT) states that the learning process can be facilitated by

For PAEE: *"Mathematics and Arts"*.

What we see is a total lack of communication between the teachers who work with this student. Each one gives a different answer.

By organizing teaching based on students' prior knowledge. In this sense, they seek to understand the importance of prior knowledge in science teaching and identify the role of prior knowledge in the process of learning meaningfully (PEREIRA; PADILHA, 2014 a; 2014 b).

They don't match up at all. And the saddest thing about all this is that the mainstream teacher is so unmotivated and uninterested in the student's learning.

The tenth question was: "Does the school help develop new methodologies in the classroom?"

PC and PAD both answered "*Sometimes*".

The PAEE said: *"Yes, the school contributes effectively and financially to this construction"*.

Farias (2008) points out that inclusion should be instituted as a form of radical, complete and systematic integration, in which schools should set out to adapt their educational systems to the special needs of their clientele, all students, not just those with disabilities. And this is clearly not happening in the case under investigation.

It's a school with a *slogan* of inclusion, but its professionals make different efforts to achieve this. The main one (a mainstream teacher) makes no effort to teach science to the ATEA.

5 FINAL CONSIDERATIONS

This study investigated whether the pedagogical practices adopted by a regular school natural science teacher contribute to the process of teaching and learning science to a student with ASD.

What was observed was that, although the mainstream science teacher has 25 years of teaching experience, her methodological practices are not adequate or sufficient to work with the student in question, leaving a great deal of responsibility to the ESA teachers and the student's class assistant teacher, showing no interest in developing the student's scientific literacy.

In view of the results found, it is necessary for future science teachers and those already working in the field to seek out new knowledge in addition to that taught at universities. Continuing teacher training is recommended so that they can work with the specific needs of their students, whatever they may be.

Despite the advances in the curriculum of undergraduate courses recommended by legislation in relation to Special Education, the results of this study show that there are numerous weaknesses in the pedagogical practice of science teachers in relation to working with students with ASD.

In this way, it is considered that interaction between teachers and students with ASD is essential; that the use of differentiated and adapted methodologies in the regular classroom is necessary for the development of scientific knowledge in students with ASD; that team lesson planning is important for the scientific and civic education of these students.

It was observed that the results of this study confirmed the initial hypothesis about the weaknesses of science teachers who work in regular education in their pedagogical practice to work with students with ASD. In this way, they do not contribute to the process of teaching and learning Science with these students insufficient teacher training, and how this interferes negatively on the learning of these students who need a truly inclusive education, we conclude that much remains to be done to ensure that scientific literacy reaches all students, including those with ASD.

REFERENCES

ABUJADI, Caio. **How an autistic person's brain works**. 2014. Available at:<https://www.youtube.com/watch?v=t7i5LzSuCCY>. Accessed on: June 2015.

ALVES, C. **Inclusive education in the regular school system - The case of the municipality of Rio de Janeiro**. 2002. Available at: <http://www.cnotinfor.pt/inclusiva/pdf/Educacao inclusiva_RJ>. Accessed on: June 2015.

ASSUMPQAO JUNIOR, Francisco B.; PIMENTEL, Ana Cristina M. Autismo infantil. **Rev. Bras. Psiquiatria**, n. 22 (Supl I), p. 37-9, 2000.

AYRES, Antonio Tadeu. **Competent Pedagogical Practice**: Expanding Teacher Knowledge. 5 ed. Petropolis, RJ: Vozes, 2011.140p.

BARDIN, Laurence. **Content analysis**. Translated by Luis Antero reto; Augusto Pinheiro. Sao Paulo: ed.70, 2011.

BEREOHFF AMP. **Autism, a multidisciplinary view**. Sao Paulo: GEPAPI, 1991.

BRAZIL. LAW NO. 12.764, OF DECEMBER 27, 2012. Establishes the National Policy for the Protection of the Rights of Persons with Autism Spectrum Disorders; and amends § 3° of art. 98 of Law n° 8.112, of December 11, 1990.

BRAZIL. Ministry of Health. Health Care Secretariat. Department of Strategic Programmatic Actions. **Guidelines for the Rehabilitation of People with Autism Spectrum Disorders**. Brasilia: Ministry of Health, 2013. 74p. Available at:
<www.saude.gov.br>. Accessed on: Nov., 2014.

BRASIL. **Política Nacional de Educação Especial na Perspectiva da Educação Inclusiva**. Brasilia: MEC/SEESP, 2008.

BRIDI, F. R. S.; FORTES, C. C.; BRIDI FILHO, C. A. Education and autism: the subtleties and possibilities of the inclusive process. In: ROTH, B. W. (Org.) **Experiencias educacionais inclusivas**: Programa de educapao inclusiva: direito a diversidade. Brasilia: Ministerio da

Educapao, Secretaria de Educapao Especial, 2006.

CAIXETA, J. E. et al. Science teacher training: experience of the subject educating with special needs at the x extension week of the university of Brasilia. **Anais...** XI CONGRESSO IBEROAMERICANO DE EXTENSION UNIVERSITARIA, November 22-25, 2011, Santa Fe, Argentina. Brasilia-Brazil: UNL, 2011. 20p.

CHASSOT, Attico. **Scientific literacy**: challenges for education. 6 ed. ijuf: UNIJUI, 2014. 368p. (Chemistry Education Collection)

CHASSOT, Attico. **Scientific literacy**: issues and challenges for education. Ijui: Unijui, 1 ed. 2000, 434 p., 2 ed. 2001,438 p.

FARIAS, Iara Maria de et al. Teacher-student interaction with autism in the context of inclusive education: analysis of the teacher's mediation pattern based on the Mediated Learning Experience Theory. **Revista Brasileira de Educapao Especial**, 2008,17p.

FURTADO, Angelo Amaro Martins; KAWAKAMI JUNIOR, Masahiko; PEREIRA Gerlany de Fatima dos Santos. Rape: methodological possibilities for teaching Science in Elementary School II. In: **Anais...** I CONGRESSO INTERNACIONAL DE FORMAQAO DE TEACHORES DO AMAPA - III FORUM DAS LICENCIATURAS DA UEAP, Macapa-AP, May 2015.

GIL-PEREZ, Daniel; VILCHES, Amparo. Education, citizenship and scientific literacy: myths and realities. **Revista Iberoamericana de Educacion**, n. 42, p. 31-53, 2006.

GILLBERG ,Christopher ; Autism spectrum disorders.In: **Anais...** Congress of Psychiatry, Belo Horizonte ,2005.

KLIN, A. Autism and Asperger syndrome: an overview. **Rev Bras Psiquiatr**. v. 28, (Supll I), p. 3-12, 2006.

LEON, Maria Josefa Guerrero. Biology in the new baccalaureate. **Alambique**, v. 36, p. 76-81, 2003.

LORENZETTI, Leonir; DELIZOICOV, Demetrio. Scientific literacy in the context of the early grades. **Ensaio - Pesquisa em Educapao em**

Ciencias. v. 3, n. 1, jun., p. 1-17, 2001.

MELLO, A. **Autismo**: guia pratico. Brasilia: CORDE, 2004.

MILAGRE, Marilene de Oliveira; SOUZA, Wagna da Silva. **A study into the integration of autistic children into mainstream education**. Serra: Escola Superior de Ensino Anisio Teixeira, 2011.42p.

OLIVEIRA, M. **Vygotsky**: Learning and Development - a socio-historical process. 4. ed. Sao Paulo: Scipione, 1997.

PASSERINO, Iliana M. Alternative communication, autism and technology: case studies from Scala. In: MIRANDA, Theresinha Guimaraes; GALVAO FILHO, Teofilo Alves (Orgs.). **Teachers and inclusive education**: training, practices and places. Salvador: EDUFBA, 2012.p. 223-246.

PASSOS, Marileni O. et al. **Fundamentals and methodologies of special education**. Curitiba: Fael, 2011.63p.

PEREIRA, G. F. S.; PADILHA, E. C. P.; COSTA, F. J. S. Immersion in the universe of the Croods animation: possibilities for teaching science. In: **Anais...** 1ST INTERNATIONAL CONGRESS OF
AMAPA TEACHERS' TRAINING AND III FORUM OF THE LICENCIATURES OF THE UNIVERSITY OF THE STATE OF AMAPA, Macapa -Ap, 2015.

PEREIRA, Gerlany de Fatima dos Santos; PADILHA, Elane Cristina Pereira. The Importance of Prior Knowledge in Science Teaching in the Light of David Ausubel's Meaningful Theory. In: **Anais...** V ENAS - NATIONAL MEETING ON SIGNIFICANT LEARNING, Belem-Pa, 2014a.

PEREIRA, Gerlany de Fatima dos Santos; PADILHA, Elane Cristina Pereira. The Role of Prior Knowledge in the Process of Learning Meaningfully: Approaches to Science Teaching. **Anais...** V ENAS - ENCONTRO NACIONAL DE APRENDIZAGEM SIGNIFICATIVA, Belem-PA, 2014b.

PEREIRA, Gerlany de Fatima dos Santos; RIBEIRO, Elinete Oliveira Raposo; FREITAS, Nadia Magalhaes da Silva. **Socio-scientific controversies in science teaching**: AGROBIO in focus. Novas Edipoes Academicas/Verlag Editora: Saarbrucken, Deutschland, 2016. 4 p.

PEREIRA, Gerlany de Fatima dos Santos; RIBEIRO, Elinete Oliveira Raposo; FREITAS, Nadia Magalhaes da Silva. **Appropriation of Scientific Knowledge**: an approach to food transgenics. Novas Edipoes Academicas/Verlag Editora: Saarbrucken, Deutschland, 2014. 104 p.

PERES DE PAULA; Adapted children's literature in inclusive education: inclusion alternatives for autistic students from a socio-community perspective. **Revista de Ciencias da Educapao**, 2014.

PERORAZIO, D.; **My famous warrior**. 1 ed, Sao Paulo: Biblioteca, 2009.

PESSOA. Nataly. **Classic Autism** . In: <http://espacoautista.blogspot.com.br/2012/10/autismo-classico.html> Accessed May 9, 2013.

PINHEIRO, Daiane; Duarte, Sabrina Maiara de Sousa. ESA teachers and regular classroom teachers: articulations, evaluations and effects on the teaching and learning of included students. **Anais...** CONGRESS 6TH SBECE AND 3RD SIENCE EDUCATION, TRANSGRESSION AND NARCISSISM. 2014.

PRAQA, Elida Tamara Prata de Oliveira. **A reflection on the inclusion of autistic students in mainstream education.** 2011. 140f. Dissertation [Master's]. Federal University of Juiz de Fora. Institute of Exact Sciences, Juiz de Fora (MG), 2011.

RIVIERE, A. Coleccion Estructura y Processos. Pensamiento, **Psicopatologia y Psiquiatria** series. Madrid: [s.n.], 2001.

SACRISTAN, Jimeno. **Unstable powers in education**. Porto Alegre: ARTMED Sul, 1999.

SADLER, Troy D. Informal Reasoning Regarding Socioscientific Issues: A Critical Review of Research. **Journal of Research in Science Teaching**, v. 41, n. 5, p. 513-536, 2004.

SANTIAGO, Sandra Alves da Silva, Myths and truths that every teacher should know - Reflections on pedagogical practice from the perspective of inclusion. **Construir *Noticias* magazine**, June 2004.

SANTOS, Manuella Teixeira; FERREIRA, Silvaney Fonseca; SANTANA,

Elizangela Barreto; PEREIRA, Gerlany de Fatima dos Santos; FREITAS, Nadia Magalhaes da Silva. Photography and science teaching: graduate students' impressions of the photography experience. In: **Anais...** IBERO-AMERICAN CONGRESS ON SCIENCE, TECHNOLOGY, INNOVATION AND EDUCATION. Buenos Aires, Argentina, November 12-14, 2014.

SOUZA, Vinicius Catao de Assis. Science teaching and its inclusive challenges: a chemistry teacher's view of school (in)difference. **Revista de Ciencias.** 2013.

WING, L. **El Autismo en ninos y adultos**: una guia para la familia. Buenos Aires: Paidos, 1998.

YIN, Robert K. **Case study**: planning and methods. Trad. Daniel Grassi. 2 ed. Porto Alegre: Bookman, 2001.

APPENDIX A - OBSERVATION SCRIPT

OBSERVATION SCRIPT - PLANNING

- What profile do teachers present in their teaching practice?
- What theoretical-methodological approach is expressed in the learning environment observed?
- Is there consistency between what was proposed in the lesson plan and what was actually done by the teacher?
- How does science teaching work in the classroom?

TEACHER-STUDENT RELATIONSHIP

- How do interpersonal relationships occur between teachers and students?
- How do teachers deal with differences in the learning pace of students, especially those with autism spectrum disorder?

METHODOLOGICAL PROCEDURES

- Does the methodology used in the classroom include practical science activities? In what way?
- Does the teacher's mediation allow for meaningful learning?
- Is the content contextualized with the students' socio-cultural reality?
- What form does pedagogical practice take?
- Are most of the activities proposed to the class individual or collective?

USE OF RESOURCES AND DEVELOPMENT

- Are the resources used properly?
- They are suitable for students with autism spectrum disorder.
- Are the resources motivating and enrich the development of the lesson?
- Are the didactic-pedagogical resources appropriate for the students, and how do these resources contribute to the learning of students with autism spectrum disorder?
- How is the student's development monitored?
- What aspects are considered for this record?
- How do those responsible monitor this development?
- What points do you consider to be positive in the teacher's pedagogical practice?
- What points do you consider to be negative in the teacher's

pedagogical practice?
• What kind of pedagogical interventions are necessary in the observed pedagogical practice?

APPENDIX B - INTERVIEW SCRIPT

Professor:

1. Did you have any preparation in your degree course to know how to teach and deal with students with autism spectrum disorders?

2. What practices have you adopted for the development and integration of students in the classroom, especially those with Autism Spectrum Disorder?

3. What is your biggest difficulty in the classroom with students with autism spectrum disorder included in mainstream classes?

4. Do you use any teaching aids to help you understand the content better? What are they?

5. Is this student managing to develop didactically in the school environment? How does this happen?

6. How is science teaching suited to this student?

7. During science lessons, do students show any interest in the content?

8. How is your relationship with the EYFS teachers, and do they give you tips on suitable methodologies?

9. What content did you teach in class that you felt was easier for the student to assimilate?

10. Does the school help develop new methodologies in the classroom?

ANNEX A – INFORMED CONSENT FORM

You are being invited to participate as a volunteer in a study,_____. After clarifying the following information, if you agree to take part in the study, please sign at the end of this document, which is in two copies. One is yours and the other is the researcher's. If you refuse, you will not be penalized in any way. If you have any questions, please contact _____ or telephone_____.

INFORMATION ABOUT THE RESEARCH:
Project title:_____.
Work supervisor:_____.
Contact telephone number (including collect calls)_____:.
The work aims to._____
And it will be held at period_____ and does not imply any risk to life and/or morals and will not cause any damage, discomfort and/or injury.
The academic_____, **as well as *the supervisor*_____**, undertake not to divulge information given by interviewees who withdraw consent to the interview throughout the course of the research; that in this case, the interviewee may withdraw consent at any time during the above-mentioned period, and when the information is kept, it will be under the complete confidentiality of the interviewee, in which case a pseudonym and/or abbreviation of the interviewee's name will be used with an indication of the location of the interview.

Researcher/student/institution

_____

ANNEX B - CONSENT FOR THE PARTICIPATION OF THE PERSON AS SUBJECT

Eu,_____,
RG / CPF/ n.°_____ ,
undersigned, agree to participate in the study entitled "**Teaching practices in science education: Autism in question**", as a subject.

I have been duly informed and clarified by the student/researcher Dayana Tavares Chaves about the research, the procedures involved, as well as the possible risks and benefits arising from my participation.

I have been assured that I can withdraw my consent at any time, without this leading to any penalty or interruption of my follow-up/assistance/treatment.

Place and date

_____

Name of subject or guardian:

Signature:

_____

We witnessed the request for consent, clarifications about the research and the subject's acceptance to participate
Witnesses (not connected to the research team):

Name

_____

Signature

Name

Signature

Additional comments

I want morebooks!

Buy your books fast and straightforward online - at one of world's fastest growing online book stores! Environmentally sound due to Print-on-Demand technologies.

Buy your books online at
www.morebooks.shop

Kaufen Sie Ihre Bücher schnell und unkompliziert online – auf einer der am schnellsten wachsenden Buchhandelsplattformen weltweit! Dank Print-On-Demand umwelt- und ressourcenschonend produziert.

Bücher schneller online kaufen
www.morebooks.shop

 info@omniscriptum.com
www.omniscriptum.com